Tommy, look up:
Up high in the Sky.
The stars above are seen twinkling
It's a beautiful gift to see the stars shining!

That's not all that catches the eye.
There is something big reflecting light.
It's the moon glowing ever so bright.
Changing its form from night to night!

The moon goes through eight phases
That we can see as the moon's glow changes.
Eight phases over about twenty-nine days.
Each phase gives the moon a new face!

Tommy, look up:
Up in the sky:
It's the Waxing Crescent Moon:
The moon is just starting to glow.
The moon's beginning to look like a half halo!

Tommy, look up.
Up in the sky.
It's the First Quarter Moon.
The moon is half glowing up high.
The moon is looking like half a pie!

Tommy, look up:
Up in the sky:
It's the waxing Gibbous moon.
Waxing means growing;
The moon is nearly fully glowing!

Tommy, look up.
Up in the sky.
It's the full moon.
The whole moon is filled with light.
The night sky is glowing and bright!

Tommy, look up:
Up in the sky:
It's the waning gibbous moon.
Waning means shrinking.
Every night the moon keeps on dimming!

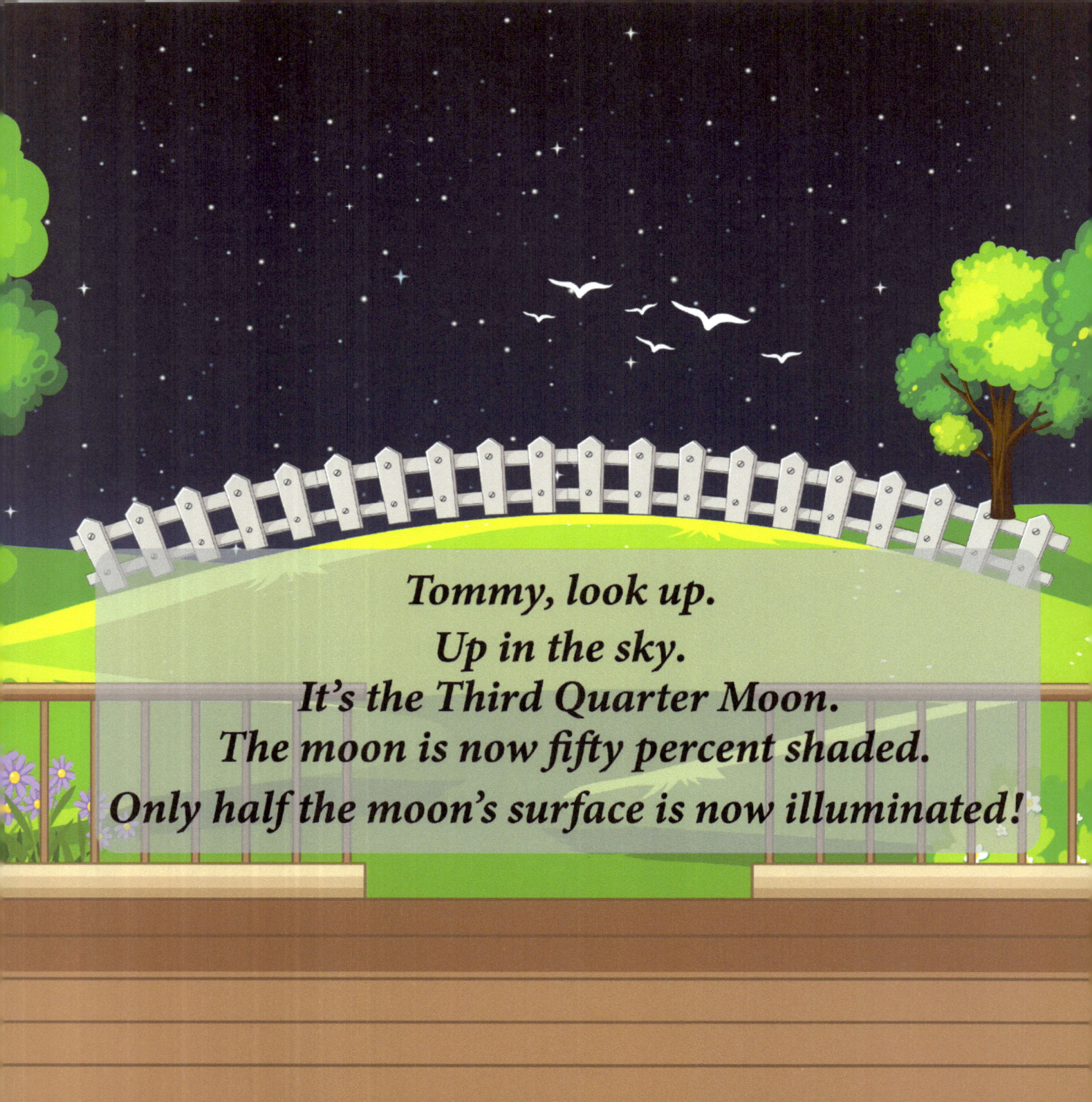
Tommy, look up.
Up in the sky.
It's the Third Quarter Moon.
The moon is now fifty percent shaded.
Only half the moon's surface is now illuminated!

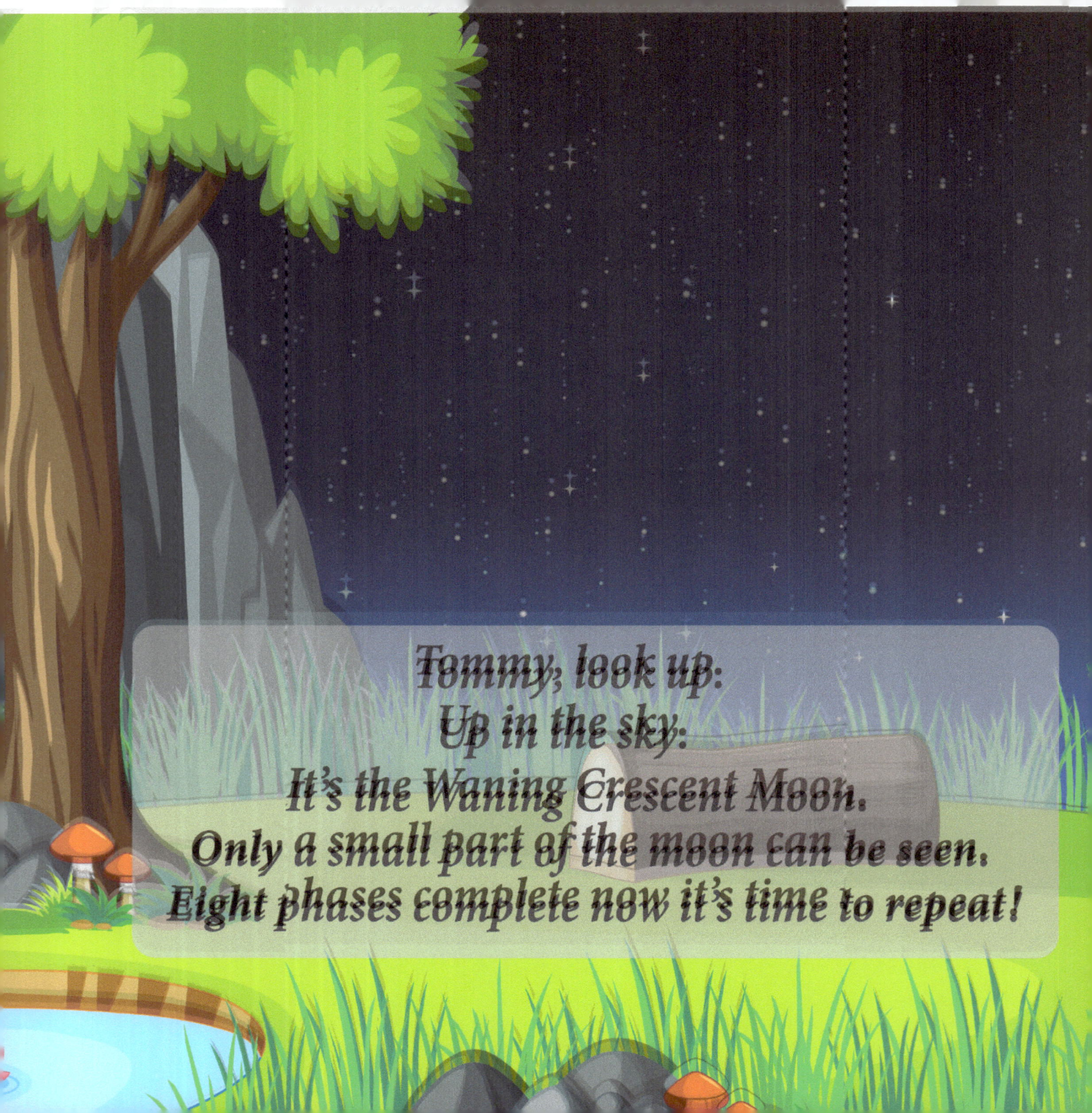

Tommy, look up:
Up in the sky:
It's the Waning Crescent Moon.
Only a small part of the moon can be seen.
Eight phases complete now it's time to repeat!

THE END

www.ingramcontent.com/pod-product-compliance
Lightning Source LLC
Chambersburg PA
CBHW051953210526
45473CB00024B/2373